ISBN 978-0-666-89343-7
PIBN 11336950

STUDIES ON THE PRODUCTION OF 2,3-BUTYLENE GLYCOL FROM WHEAT BY AEROBACILLUS POLYMYXA

Albert William Jackson

Department of Field Crops

University of Alberta

STUDIES ON THE PRODUCTION OF 2,3-BUTYLENE GLYCOL

FROM WHEAT BY *AEROBACILLUS POLYMYXA*

Albert William Jackson

Department of Field Crops

A THESIS

submitted to the University of Alberta in

partial fulfilment of the requirements

for the degree of

MASTER OF SCIENCE

Edmonton, Alberta

April, 1944

TABLE OF CONTENTS

	Page
Introduction	1
Previous work at the University of Alberta	3
Objects of the present investigations	4
Pasteurization in the isolation and improvement of _A. polymyxa_	5
Effect of pasteurization upon the glycol-yielding ability of _A. polymyxa_ strains	5
Introduction	5
Methods	5
Results	7
Strain U.A.500	7
Strain U.A.350	8
Application of pasteurization to the isolation of _A. polymyxa_ strains	11
Introduction	11
Method	12
Results	13
Comparative glycol yields obtained from various _A. polymyxa_ strains	14
Introduction	14
Method	14
Results	15
Discussion	17

TABLE OF CONTENTS (Continued)

Page

Morphological, biochemical and cultural characteristics of A. polymyxa 18

Introduction 18

Materials and method 21

Results 22

Morphology 22

Cultural characteristics 22

Gelatin liquefaction 25

Potato rotting 25

Acetymethylcarbinol formation 25

Nitrate reduction 25

Temperature relations 27

Discussion 27

Growth factor requirements of A. polymyxa 28

Method 29

Results 31

The nitrogen requirements of A. polymyxa 33

Introduction 33

Nitrogen requirements of A. polymyxa for the fermentation of starch 35

Introduction 35

Materials and methods 36

Description of wheat-flour fractions used as supplements to starch 36

TABLE OF CONTENTS (Continued)

	Page
Liquid gluten	36
Starch wash-liquor	36
Results	39
The amino acid requirements of *A. polymyxa*	41
Introduction	41
Method	42
Results	43
Amino acid requirements of *A. polymyxa* in the presence of biotin, thiamin, nicotinic acid, pantothenic acid and inositol	43
Amino acid requirements in the presence of biotin and thiamin	45
Inorganic nitrogen and urea as nitrogen sources for *A. polymyxa*	46
Introduction	46
Method	47
Results	47
Training of *A. polymyxa* to utilize ammonium sulfate	48
Introduction	48
Method	49
Results	50

TABLE OF CONTENTS (Continued)

	Page
The effect of liquid gluten on the glycol and ethanol yields of A. polymyxa starch fermentations	52
Introduction	52
Methods	52
Results	54
The effect of ammonium sulfate at various gluten concentrations on glycol yields from starch fermentations	54
Introduction	54
Method	55
Results	55
Discussion	56
Summary	58
Acknowledgements	61
Literature cited	62

STUDIES ON THE PRODUCTION OF 2,3-BUTYLENE GLYCOL FROM WHEAT BY <u>AEROBACILLUS POLYMYXA</u>

Albert William Jackson

INTRODUCTION

Owing to the loss during the present war of the main producing areas of natural rubber the United Nations were forced to seek other sources of supply. Two general methods for overcoming the shortage were suggested. These were the production of synthetic rubber and the utilization of the latex of certain quick-growing plants.

The main processes for the production of synthetic rubber require butadiene as a basic substance. Butadiene can be produced from a variety of compounds - e.g. 2,3-butylene glycol, ethyl alcohol and some of the constituents of natural gas. It is also obtained as a by-product in the cracking and refining of petroleum (9).

The study of the production of 2,3-butylene glycol was undertaken mainly because of its possible conversion to butadiene. In addition the "laevo" form of 2,3-butylene glycol in water solutions has been shown to have definite possibilities as an anti-freeze (10).

Certain species of bacteria produce 2,3-butylene

glycol (referred to hereafter as glycol) as a fermentation product. Among these species are Aerobacillus polymyxa (Prazmowski) Donker producing the "laevo" form, and Aerobacter aerogenes producing a mixture of "dextro" and "meso" forms in relatively large amounts by the fermentation of carbohydrates. The former is able to ferment starch. Since cereals contain large amounts of starch the Aerobacillus fermentation has been suggested as a means of producing glycol and of reducing the wheat surplus.

The Aerobacter fermentation on the other hand requires a simple carbohydrate in addition to continued aeration during fermentation. It is also particularly sensitive to contamination by other organisms. These are handicaps to the adaptation of this fermentation to the commercial production of glycol.

Organisms of the Aerobacillus type, in addition to possessing starch hydrolyzing enzymes, do not require aeration during fermentation. Because of this there is less opportunity for contamination in the Aerobacillus than in the Aerobacter fermentation.

Early in 1942 the Department of Field Crops of the University of Alberta, co-operating with the National Research Council Laboratory, Ottawa, began a war emergency research project designed to study the possibility of utilizing wheat in the production of glycol using Aerobacillus polymyxa.

would [illegible] hazardous [illegible]

[illegible]

[illegible] form, and

[illegible]

[illegible]

[illegible] of carbohydrate. [illegible]

[illegible]. Since [illegible]

[illegible]

made of [illegible]

The [illegible]

requires a [illegible]

[illegible]

sensitive [illegible]

and [illegible]

[illegible] of glucose.

[illegible]

[illegible]

[illegible]

[illegible] fermentation.

[illegible] Department of [illegible]

[illegible]

[illegible] Laboratory, [illegible]

[illegible] the possibility of

[illegible] production [illegible]

[illegible]

PREVIOUS WORK AT THE UNIVERSITY OF ALBERTA

Not all phases of our work on the *Aerobacillus* fermentation are included in the present report. The early work conducted in this laboratory dealt with the following:

1. The isolation of active starch-hydrolyzing and glycol-producing strains of *A. polymyxa* from different sources and the improvement of these strains by re-selection from the original isolates;

2. The determination of the suitability of various wheat fractions as nutritional supplements for *A. polymyxa* in starch fermentations;

3. The perfection of methods for the comparative fermentation of a large number of grain samples and the determination of the relative values of different grades, varieties, etc., of wheat for glycol production.

A major part of our previous work was devoted to the third sub-project.

Results of work referred to in sections 1 and 2 above are to be found in progress reports submitted to the National Research Council by this laboratory (7). A description of the equipment and the methods devised to carry out the fermentations referred to in Section III may

be found in the University of Alberta Report (17) Section II and the glycol and ethanol yields obtained from the various grain samples in the latter (17), Section I.

OBJECTS OF THE PRESENT INVESTIGATIONS

The objects of the present investigations were as follows:

1. (a) To continue the work of endeavoring to obtain strains of A. polymyxa having superior glycol-yielding ability to those now available by the improvement of existing strains and by further isolation from local source material;

   (b) To study the cultural characteristics of the higher-yielding strains, and to determine if they belong to a single group or if they have any property or properties in common by means of which they may be quickly identified;
2. To study the nutrition of A. polymyxa, especially its nitrogen and growth factor requirements with the aim of using the Aerobacillus fermentation to produce glycol from more or less pure wheat starch.

PASTEURIZATION IN THE ISOLATION AND IMPROVEMENT OF *A. POLYMYXA*

Effect of Pasteurization upon the Glycol-yielding Ability of *A. polymyxa* Strains

Introduction

The spore-forming bacteria (*Clostridium acetobutylicum*), used in the commercial production of acetone and butanol, were observed by Weyer and Rettger (21) to become sluggish in action and to produce reduced yields of solvents after being transferred on the usual culture medium for some time. It was found (21) that pasteurizing such cultures increased their activity. This treatment removed the non-spore-bearing forms which are poor solvent-producers. The strains producing the highest yields of solvents were the most heat-resistant spore-bearing isolates.

Similarly in the present work reduction in the glycol yields obtained from various strains of *A. polymyxa* had been noted after continued transfer on a peptone-wheat-agar medium. It was decided to attempt the activation of these organisms by pasteurization, using the following methods.

Methods

Spore-containing culture was transferred from agar

slants to sterile 0.9 percent NaCl solution in small diameter (14 mm.) Pyrex test-tubes. These were heated in a water bath at temperatures of 65°, 75° or 85° C for 10 minutes and then cooled. A non-pasteurized control was held at room temperature.

The effect of heat upon the material prepared in the manner described above was studied by two methods:

(a) The spore-suspension was used to inoculate duplicate 150 c.c. lots of 8 percent wheat mash which were then fermented and the resulting mixtures analyzed for glycol and reducing sugar;

(b) One loopful of the spore-suspension was added to 10 c.c. of neutral red agar containing starch. This was then poured into Petri plates. Aerobacilli usually absorb the red dye (15). It was observed that the addition of starch to this medium gave it a further advantage in that active starch hydrolyzers could be selected from among the Aerobacilli which did not absorb the dye.

After incubation for 48 hours to 72 hours at 30° C, colonies judged to be Aerobacillus colonies were picked from the plates and transferred to test-tube lots of sterile 6 percent wheat mash. These were placed in an oven held at 30° C. At the end of 48 hours' incubation the contents of actively fermenting tubes were used to inoculate 8 percent wheat mashes which were treated as in method (a).

The wheat mashes were prepared by cooking 1 to 1.5 litre batches of mash in a hot water bath at the temperature of boiling water and cooking was continued for 30 minutes with constant stirring. At this time $CaCO_3$ at the rate of 0.5 gm. per 100 c.c. water in the mash was added and uniformly dispersed by further stirring. The cooked mash was dispensed in 150 c.c. lots into 250 c.c. flasks which were then bunged and autoclaved at 15 lb. steam pressure for 30 minutes.

After cooling the mash, the inoculum prepared in (a) and (b) was added and dispersed by shaking the flasks. The inoculated flasks were placed in a cabinet held at 35° C for 8 to 12 hours, then removed to a cabinet held at 30° C for the remainder of the fermentation period. Analyses of the spent mashes were made after 72 hours' fermentation.

Results

Strain U.A.500

Results obtained in two tests on 8 percent wheat mash using method (a) and organism U.A.500 are given in Table I.

Cultures from the spores resistant to 85° C produced higher yields of glycol than those from spores subjected to pasteurization at 65° C or those from unpasteurized controls. The tests show that consistent increases in glycol yield can be obtained by this procedure.

There is little or no trend in the ethanol yields

TABLE I

Effect of pasteurizing upon the glycol yield of <u>A. polymyxa</u> strain U.A.500 using method (a)

	Temperature of Pasteurization								
	Check (not Past.)			65° C			85° C		
Test No.	% Glycol	% Etoh	% Sugar	% Glycol	% Etoh	% Sugar	% Glycol	% Etoh	% Sugar
1.	1.04	0.66	0.25	1.09	0.67	0.20	1.17	0.69	0.21
	1.01	0.68	0.23	1.07	0.70	0.21	1.18	0.73	0.21
Mean	1.02	0.67	0.24	1.08	0.69	0.20	1.17	0.71	0.21
2.	1.10	0.64	0.28	1.18	0.67	0.27	1.28	0.76	0.19
	1.11	0.68	----	1.16	0.66	0.32	1.29	0.73	0.27
Mean	1.10	0.66	0.28	1.17	0.66	0.29	1.28	0.74	0.23

and no differences in the residual sugar which can be accounted for by pasteurization of the cultures.

<u>Strain U.A.350</u>

A summary of the results obtained using methods (a) and (b) with strain U.A.350 is given in Table II.

Strain U.A.350 differs from U.A.500 in that it produces optimum yields of glycol after pasteurization at 65° C. This trend was evident in both tests where method (b) was used.

Increasing the temperature of pasteurization to

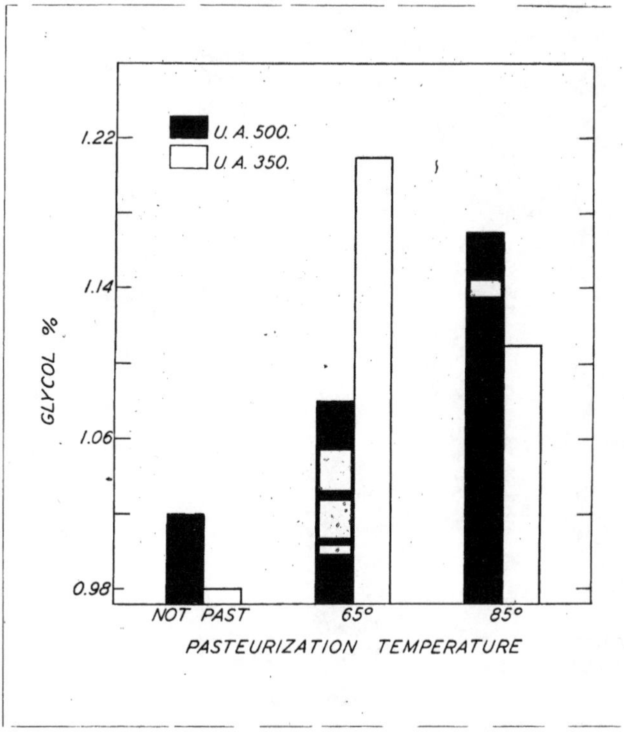

Figure 1

Effect of pasteurization upon the glycol-yielding ability of strains U.A.500 and U.A.350

TABLE II

Effect of pasteurizing upon the glycol yield of A. polymyxa strain U.A.350 using methods (a) and (b)

		Temperature of Pasteurization							
		Check		65° C		75° C		85° C	
Test No.	Method	% Sugar*	% Glycol	% Sugar	% Glycol	% Sugar	% Glycol	% Sugar	% Glycol
1.	(a)	0.28	0.97	0.40	1.18	0.19	1.20	0.27	1.15
		0.25	0.99	0.09	1.24	0.27	1.23	0.32	1.08
	Mean	0.26	0.98	0.25	1.21	0.23	1.21	0.29	1.11
	(b)	0.14	1.09	0.13	1.24	0.11	1.08	0.08	1.12
		0.11	1.03	0.09	1.22	0.21	0.96	0.13	1.00
		0.14	1.06	0.10	1.14	0.10	0.97	0.15	1.01
		0.12	1.08	0.12	1.28	0.09	0.99	0.28	0.99
	Mean	0.13	1.06	0.11	1.22	0.13	1.00	0.16	1.03
2.	(a)	0.09	1.03	0.01	1.27	----	----	0.05	1.11
		0.21	1.04	0.13	1.20	----	----	0.11	1.14
	Mean	0.15	1.03	0.07	1.23	----	----	0.08	1.12
	(b)	0.28	1.14	0.17	1.23	----	----	0.23	1.08
		0.13	1.16	0.06	1.32	----	----	0.14	1.14
		0.20	0.93	0.18	1.24	----	----	0.16	1.10
		0.39	1.12	----	----	----	----	0.26	1.09
		0.19	1.15	----	----	----	----	0.29	1.18
	Mean	0.22	1.10	0.13	1.26	----	----	0.22	1.12

* Reducing Sugar.

85° C reduced the glycol yields obtained with U.A.350. This trend was observed in all tests.

When the above experiments were concluded work was begun using 15 percent instead of 8 percent wheat

mashes. Results obtained using the higher mash concentrations showed no consistent trends, although there was some tendency for organisms to respond to pasteurization at either 65° or 85° C.

The Application of Pasteurization to the Isolation of *A. polymyxa* Strains

Introduction

The results just reported indicate that pasteurization of a culture increases the glycol yield obtained from the subsequent fermentation of 8 percent wheat mashes.

Stanier and co-workers (15) included pasteurization of source material in their method of isolating high glycol-yielding strains of *A. polymyxa*. They concluded that pasteurization was an essential part of the procedure on the basis of the high percentage of active isolates which they were able to obtain by its inclusion. However no direct comparisons were made of the number or yielding ability of isolates obtained from pasteurized and non-pasteurized source material, nor was the effect of various temperatures of pasteurization reported.

On the strength of the trends recorded in the previous section and the observations of Stanier *et al* (15) it was decided to study the value of pasteurization in

isolating high-glycol-yielding strains of A. polymyxa from soil and other sources.

Method

With the following modifications the method used was that described as method (b) in the previous section.

Material for pasteurization was prepared by adding 1 c.c. of a 1 to 100 dilution of source material in water to 5 c.c. of sterile water in test-tubes.

Pasteurization of these tubes was carried out at the following temperatures: 65°, 80° and 96° C. A non-heated control was included in the test.

The 8 percent wheat mashes were replaced by 15 percent mashes which were prepared by a method somewhat similar to that described for the preparation of the less concentrated ones. However, in addition to the increase in concentration, the following changes were made in the preparation of the mashes. Owing to the almost solid nature of the cooked, non-liquefied 15 percent wheat mashes, pre-liquefaction was found necessary. Malt extract was used to liquefy the starch. Special equipment which was designed and constructed for the preparation of uniform mashes for routine fermentations was used in these studies. A full description of the methods and equipment used in the preparation of 15 percent wheat mashes is to be found elsewhere. (17)

Cultures were preserved by making loop transfers

from actively fermenting wheat tubes to agar slants.

Results

The results obtained in three consecutive tests on different source materials are summarized in Table III.

TABLE III

The value of pasteurization in the selection of high glycol-yielding strains of A. polymyxa from source material

Test No.	Source Material		Check	65°	80°	96°	96° for 20 min.
			Temperature of Pasteurization				
1.	Decayed wood	A*	0	16	27	27	-----
		B*	0	1.55	1.71	1.76	-----
		C*	0	1.92	1.85	1.87	-----
2.	Soil	A	16	25	0	36	20
		B	2.10	2.18	0	2.14	2.15
		C	2.10	2.28	0	2.18	2.15
3.	Decayed wood	A	45	50	-----	25	75
		B	2.28	2.30	-----	2.19	2.25
		C	2.34	2.32	-----	2.19	2.46

* A - Percentage of active glycol producers among those selected from the plates.
B - Average yield of active isolates (% glycol).
C - Yield of highest yielding isolate (% glycol).

Pasteurization of source material is shown to be of considerable value in isolating active, starch-fermenting, high-glycol-yielding strains of A. polymyxa.

In two tests out of three, pasteurizing at 65° C for 10 minutes was most satisfactory from the view-point of the highest glycol-yielding strain selected. In the third test pasteurizing at 96° C for 20 minutes was most satisfactory from this viewpoint.

The percentage of active glycol producers among those selected from the dilution plates is generally higher in the pasteurized series than in the controls. However this trend varied considerably in different tests.

Comparative Glycol Yields Obtained from Various *A. polymyxa* Strains

Introduction

Part of the work undertaken was that of isolating further high-glycol-yielding strains of *A. polymyxa*. The glycol yields of some of those obtained are given in Table IV. Two National Research Council isolates are included for comparative purposes. One (C42(2)) is a phage-resistant culture, and the other (C47(2)) is one of the better glycol-yielding strains in the N.R.C. collection made available for this work.

Method

Comparative yields were obtained from the fermen-

tation of 15 percent wheat mashes prepared by the method described in the previous section.

Inoculum was prepared as follows: A loopful of bacteria from a slant culture was transferred to 10 c.c. of sterile water in a test tube. Following dispersion of the cells in the water by shaking the tube, the contents were transferred to 100 c.c. of 15 percent wheat mash. After incubation at 35° C for 24 hours, 5 c.c. of the actively fermenting mixture were used to inoculate 150 c.c. of the 15 percent wheat mash.

Results

The comparative glycol yields from fermentation of 15 percent wheat mashes by some of the better isolates obtained in this work and two of the N.R.C. isolates are given in Table IV.

Glycol yields obtained from the U.A. strains listed show that by the use of the pasteurization method described earlier it is possible to isolate comparatively high-glycol-yielding A. polymyxa strains from soil and rotting wood.

The yield of the isolates obtained in this work compares favorably with that of the best glycol-yielding strains among the N.R.C. isolates.

TABLE IV

Comparative yields of glycol from C.* and U.A.* strains of A. polymyxa on 15% wheat mash

Strain used	Glycol, %	Mean	Pasteurization temperature at isolation	Source
C42(2)	2.15 2.17	2.16		
C47(2)	2.42 2.51	2.46		
U.A.634	2.55 2.43	2.49	65° C	Soil
U.A.641	2.45 2.37	2.41	96° C	Soil
U.A.647	2.39 2.26	2.32	65° C	Soil
U.A.648	2.39 2.30	2.34	Check (not past.)	Decayed wood
U.A.649	2.46 2.40	2.43	65° C	Soil
U.A.650	2.49 2.43	2.46	96° C (20 min.)	Decayed wood
U.A.651	2.31 2.34	2.32	65° C	Soil
U.A.652	2.34 2.48	2.41	65° C	Soil
U.A.653	2.33 2.42	2.38	65° C	Soil

* C. - National Research Council
U.A. - University of Alberta

NOTE: Glycol yields are single determinations on duplicate fermentations.

Discussion

The value of pasteurization in improving the glycol-yielding ability of isolates and as apart of the procedure for isolating high-glycol-yielding strains from source material has been studied.

The results of experiments dealing with pasteurization of sources of bacteria show that no one temperature is consistently more effective than another in the isolation of high-yielding strains. The effect of pasteurization upon the strains U.A.350 and U.A.500 may help to explain this lack of consistency. Larger yields of glycol were obtained in the subsequent fermentation where cultures of U.A.350 were heated to 65° C rather than to 85° C, or where no heat treatment was applied. In similar tests strain U.A.500 produced highest yields of glycol after pasteurization at 85° C. The isolation of higher-glycol-yielding strains from material pasteurized for 10 minutes at 65° C rather than at 96° C in the first two tests may indicate that strains resembling U.A.350 were predominant in the source. On the other hand the superiority of material heated to 96° C for 20 minutes, as observed in the third test, may be due to the presence of strains resembling U.A.500.

MORPHOLOGICAL, BIOCHEMICAL AND CULTURAL CHARACTERISTICS OF A. POLYMYXA

Introduction

After drawing attention to the confusion which existed concerning the identity and differentiation of the species in the genus Aerobacillus, Porter, McCleskey and Levine (12) made a study of the morphology, biochemistry and cultural characteristics of over eighty strains of this organism.

They concluded that the facultative, spore-bearing bacteria which ferment carbohydrates with the production of acid and gas and possess the following characteristics may be classified as Aerobacillus polymyxa:

Rod-shaped, spore-bearing cells, Gram negative in young cultures;

No spores produced on glucose agar;

Produce moderate, spreading, transparent growth or raised slimy colonies on glucose agar;

Colonies may be round or amoeboid, convex or flat;

May or may not liquefy gelatin;

Many reduce cooked potato to a soft pulp in 48 hours at 37° C;

Grow well at 20° but slowly, if at all, at 42° C;

Produce acetylmethylcarbinol;

Reduce nitrates to nitrites;

Produce neither acid nor gas from rhamnose or sorbitol.

Stanier et al (15) studied the morphology and the biochemical and cultural characteristics of a number of A. polymyxa strains which produce relatively high yields of glycol in wheat fermentations. Their description agrees generally with that of Porter et al (12) with the following exceptions:

Rhamnose was fermented by about half of their isolates, and strains which had been carried in pure culture for some time showed a tendency to be non-spore-forming.

They noted that colonies of various strains of A. polymyxa differed with respect to form, consistency, surface and elevation. Furthermore good differentiation with respect to colony form and dye adsorption was obtained on neutral-red agar.

Upon analyzing the various types observable on neutral-red agar it was found that they fell into five main groups. The "Classification of Aerobacillus Variants" as proposed by Stanier et al (15) is as follows:

I Colonies flat, usually small and inconspicuous with matt surface. On neutral red agar colored very pale pink. Consistency butyrous, easily emulsified.

II Small heterogeneous group, colonies large or small mucoid, smooth or rough all character-

ized by the fact that they do not absorb color from neutral red agar.

III Colonies very opaque, milky, raised and entire, usually of soft consistency but not easily emulsified. Light pink color on neutral red agar. Two sub-groups.

III A Smooth

III B Rough

IV Colonies smooth and glistening, some convex, others raised and flat on top, and yet others have a raised edge and flat sunken centre. All absorb dye from neutral red agar with great intensity and become dark red. Four sub-groups.

IV A Smooth with raised edges and sunken centre.

IV B Convex and very smooth with a tendency to be mucoid.

IV C Slightly raised with flat surface, smooth but never mucoid.

IV D Miscellaneous unclassified forms of IV.

V Colonies convex, rough and glistening with dark red color on neutral red agar. Two sub-groups.

V A Colonies secreting a tough, tenacious slime.

V B Colonies secreting a soft slime.

In the present studies several characteristics exhibited by a number of <u>A. polymyxa</u> isolates obtained here and elsewhere were observed and recorded with the idea in mind that high-glycol-yielding strains may have some distinguishing characteristic in common which might be used in their identification and isolation.

Materials and Method

Cultures were carried on a wheat-peptone-agar (W.P.A.) medium developed during the course of this work and found suitable for the purpose. This medium contains the following substances:

Peptone	3 grams
Agar	17 "
Finely ground whole wheat	(see below)
Distilled water	1000 c.c.

These substances, exclusive of the wheat, were heated in an autoclave at 15 lb. steam pressure for 2 to 3 minutes, then dispensed in 8 c.c. lots into test-tubes each containing approximately 0.5 grams of ground wheat. The tubes were bunged, autoclaved at 15 lb. steam pressure for 30 minutes, then slanted and cooled.

Neutral red agar similar to that used by Stanier and co-workers (15) and utilized in these studies has the following composition:

Peptone	10 grams
Starch	20 "
Yeast extract	5 "
Neutral red agar	0.05 "
Agar	15 "
Water	1000 c.c.

Potatoes were prepared by surface sterilization with 1:1000 $HgCl_2$ solution for 2 minutes followed by thorough

rinsing in distilled water. Slices were placed in sterile Petri plates containing moist filter paper and after surface inoculation with bacteria were incubated at 30^0 C for 72 hours.

The following strains of <u>A. polymyxa</u> were compared with respect to all characters studied: U.A.634, U.A.641, U.A.643, U.A.650, U.A.651 and U.A.652. Strains C42(2), C47(2) and C56(2) were similarly compared except with respect to colony characters. Those which were compared with respect to colony characters include all the U.A. strains obtained during the course of the pasteurization studies reported in a previous section. These include the U.A. strains mentioned above.

Results

Morphology

Cultures of all strains studied contained rod-shaped cells of varying length.

Some spores were formed on the wheat-peptone-agar slants but not on potato-dextrose-agar. Spores were located terminally.

Young cultures (18-24 hr.) growing on W.P.A. were Gram negative.

Cultural Characteristics

All isolates obtained in the course of the pasteur-

ization studies were placed in the appropriate groups according to the "Classification of Aerobacillus Variants" referred to in the introduction to this section. A summary of the distribution of the isolates among the various groups is given in Table V.

TABLE V

Distribution of isolates based upon colony form, according to the "Classification of Aerobacillus variants"

	Percentage of isolates found in each group											
Kind of isolates	I	II	IIIA	IIIB	IIIC	IVA	IVB	IVC	IVD	VA	VB	Misc.
All active isolates	14.3	8.6	8.6	37			8.6	2.8		17.1		2.8
Highest yielding isolates from the different temperature groups (see Table III)	9.1	18.2	9.1	45.4			9.1			9.1		
The six best-yielding isolates	16.7	16.7		49.9			16.7					

The summary shows that most of the active isolates obtained during the course of the pasteurization studies belong to group IIIB.

Approximately 45 percent of the highest-yielding isolates from the different temperature groups also belong to this group.

Furthermore half of the highest-glycol-yielding

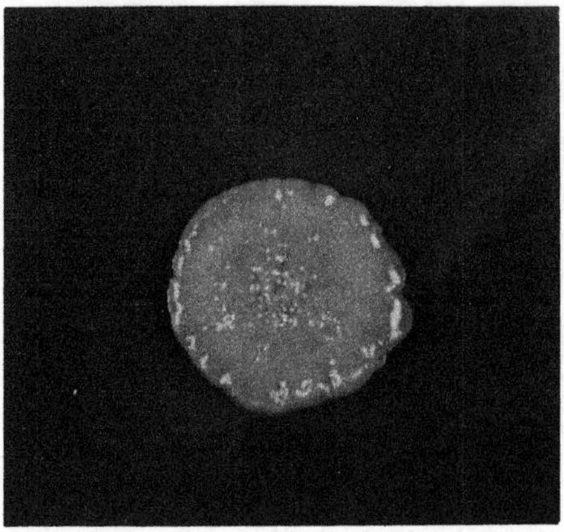

Figure 2

A typical colony of U.A.634
(Group III B)

isolates obtained in this study were classified as belonging to this group.

A typical colony is shown in Figure 2.

Gelatin liquefaction

All strains tested with the exception of C47(2) liquefied gelatin.

Potato rotting

All strains tested rotted uncooked potato slices reducing them to a pulp in less than 72 hours at 30° C. There was considerable variation in the color of the resulting pulp (see Figure 3), but as far as this has been studied these variations did not appear consistent enough to be depended upon in differentiating between high- and low-yielding isolates.

Acetylmethylcarbinol formation

All strains tested produced acetylmethylcarbinol when grown in wheat mashes as well as in the regular peptone-glucose medium. This substance was detected by the method of O'Meara (11).

Nitrate reduction

In nutrient broth containing potassium nitrate

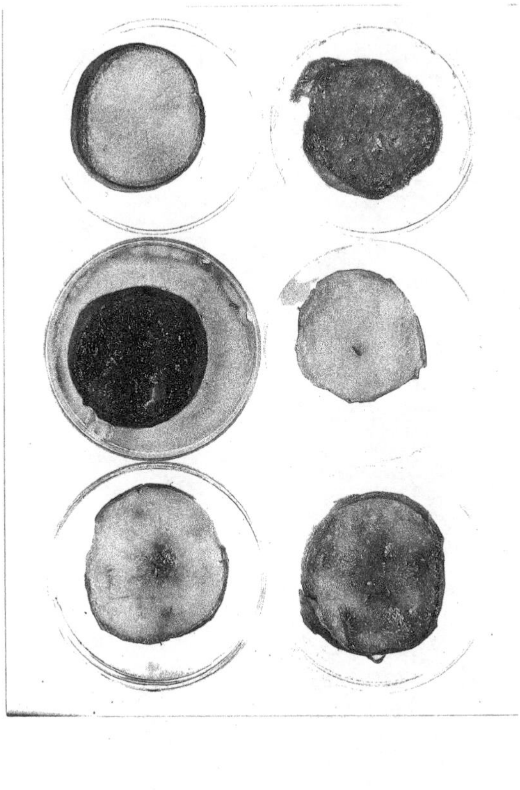

Figure 3

Rotting of whole potato slices by *A. polymyxa* strains

Left to right:	Ck.	C56(2)
	U.A.650	U.A.634
	U.A.647	U.A.641

all strains reduced the nitrate to nitrite.

Temperature relations

None of the strains tested grew on W.P.A. medium at 43° to 45° C in 48 hours. However, after incubation for 96 hours good growth of strains U.A.651 and U.A.653, slight growth of some others and no growth of C47(2) was observed. Again no differentiation between the high- and low-glycol-producing members of the group could be detected.

Discussion

A study of some morphological, biochemical and cultural characteristics of a number of strains of A. polymyxa yielded information which may be of use in further work.

The depression of spore-formation by glucose which is reported in this paper and was also observed by Porter(12) and co-workers is of importance in pasteurization studies.

Some evidence is presented to indicate that high-glycol-yielding strains of A. polymyxa may be isolated from neutral red agar plates by selecting from the colonies which have the characteristics of group IIIB. However, further study will be necessary before this method can be applied with certainty.

GROWTH FACTOR REQUIREMENTS OF A. POLYMYXA

The growth factor requirements of a number of strains of A. polymyxa have been determined by Katznelson (8) who showed that biotin was the only growth factor essential for all strains studied. Thiamin added in addition to biotin, produced irregular increases in the growth of these strains, which were divided into three groups on the basis of their reaction to this factor. Some were stimulated, others inhibited by thiamin. One strain was indifferent to this substance. Panthothenic and nicotinic acid, inositol, riboflavin and pyridoxine were not essential and produced no regular effects upon growth.

Since nothing was known concerning the growth factor requirements of several of the isolates included in the present study it was thought important to obtain this information. This was considered especially necessary in view of the fact that it was proposed to investigate pure starch as a substrate for glycol production.

Five strains were selected from among the isolates being used in the fermentation studies. These strains differed in several important characteristics such as phage resistance, glycol-yielding ability and $CaCO_3$ requirement. Strain C42(2) was included for its phage resistance, C47(2), U.A.634, U.A.641 for high-glycol-yielding ability and C56(2)

for low $CaCO_3$ requirement.

Method

The method used was essentially that employed by Katznelson (8). This consisted in growing the organisms in a chemically defined substrate containing the necessary organic and inorganic substances.

All glassware coming into contact with the medium was carefully cleaned with a dichromate cleaning solution and distilled water was used throughout the experiments.

The basal medium used was that developed by Katznelson (8). The constituents are as follows:

K_2HPO_4	1.0 grams	$CaCl_2$	0.1 grams
KH_2PO_4	1.0 "	$FeSO_4.7H_2O$	0.01 "
$MgSO_4.7H_2O$	0.2 "	$MnSO_4.4H_2O$	0.01 "
NaCl	0.1 "	$ZnSO_4$	0.1 "

Distilled water - 1000 ml.

Heat to boiling and filter

Glucose - 5 grams

pH adjusted to 6.8 with 1.0 N NaOH

Double strength medium was prepared and to it was added glutamic acid to give a concentration of 0.05 gm. per litre in the final medium. Growth factors were added as described in detail later. Distilled water was then used to bring the solution to the desired strength. The medium was dispensed in 10 c.c. lots into clean tubes which

were then bunged and autoclaved at 15 lb. steam pressure for 20 minutes.

Inoculum was prepared by suspending the bacteria in 10 c.c. of sterile 0.9 percent NaCl solution in a test tube. One 3 mm. loopful of bacteria, obtained from a 24-hour-old agar-slant culture, was suspended in the saline solution by shaking. One similar-sized loopful of the suspension was used to inoculate each tube of medium.

After 72 hours' incubation at 35° C, growth was recorded by a method similar to that employed by West and Lochhead (20) and by Katznelson (8). This consisted of rating the growth by observation of the clouding that occurred in the medium, using a scale of cloudiness ranging from 0 to 5.

When the effect of growth factors other than biotin was being studied (8) biotin was added at a uniform concentration of 0.3 micrograms per litre of medium. This concentration was used in all the work reported here. The other growth factors were added in amounts found satisfactory by Katznelson (8). These factors were added as follows:

Thiamin	200 micrograms per litre
Pantothenic acid	200 " " "
Nicotinic acid	200 " " "
Inositol	50 milligrams " "

Biotin was supplied as a concentrate containing 20 micrograms per c.c., thiamin as the chloride, and

pantothenic acid as the calcium salt.

Results

A summary of the average results obtained in two experiments conducted in triplicate is given in Table VI.

Biotin was found essential for growth of all strains used.

The presence of thiamin in addition to biotin increased the growth of four out of five strains over that obtained with biotin alone.

Each of the following, namely pantothenic acid, nicotinic acid and inositol, when used with biotin, had an effect similar to but less marked than that produced by a combination of thiamin and biotin.

The addition of pantothenic acid, nicotinic acid and inositol singly to a medium containing biotin and thiamin produced uniformly heavy growth of bacteria, generally in excess of that obtained in the presence of biotin and any other single growth factor.

No single growth factor, nor any pair of growth factors tested produced growth in the absence of biotin.

TABLE VI

Growth factor requirements of different strains of A. polymyxa

Basal + glutamic acid and the following growth factors:	Growth rating of strains used				
	42(2)	47(2)	56(2)	634	641
Check (no growth factors)	0	0	0	0	0
Biotin	2	2	4	1	3
" + Thiamin	5	5	4	4	5
" + Pant. acid	4	tr+	2	4	4
" + Nicotinic acid	4	2	4	4	4
" + Inositol	3	2	5	4	4
Biotin + Thiamin + Pant. acid	5	5	5	5	5
" + " + Nic. acid	5	5	5	5	5
" + " + Inositol	4	5	5	5	5
Thiamin + Pantothenic acid	0	0	0	0	0
" + Nicotinic acid	0	0	0	0	0
" + Inositol	0	0	0	0	0
Pantothenic acid + Nicotinic acid	0	0	0	0	0
" " + Inositol	0	0	0	0	0
Nicotinic acid + Inositol	0	0	0	0	0
Thiamin	0	0	0	0	0
Nicotinic acid	0	0	0	0	0
Pantothenic acid	0	0	0	0	0
Inositol	0	0	0	0	0

+ Trace

THE NITROGEN REQUIREMENTS OF A. POLYMYXA

Introduction

The use of whole wheat as a source of fermentable carbohydrate for the Aerobacillus fermentation did not appear to be entirely satisfactory from either a technical or an economic viewpoint. Technically the chief difficulty centered about the removal of the non-fermentable soluble and insoluble materials from the beer and the drying of this residue. Furthermore the degradation of the unused portion of the wheat protein to a by-product of questionable food value for livestock appeared to be a distinct economic disadvantage.

It was under these conditions that early in our studies of A. polymyxa we were assigned the task of developing a fermentation process for the production of 2,3-butylene glycol from pure wheat starch.

The replacement of gluten with a soluble nitrogenous substance in the fermentation mash would obviously overcome the main technical difficulties. The practicability of such a procedure depends to a large extent upon the outcome of studies on the nutrition of the organism, with special emphasis on the nitrogen requirement. The availability and cost of the alternate

nitrogen source must also be considered.

There is little published information on the nitrogen requirements of _A. polymyxa_. Tilden and Hudson (16) found a starch peptone medium more suitable for the growth of _A. polymyxa_ and production of amylase than a mineral starch medium. Katznelson (8) demonstrated that peptone, tryptone and yeast extract were suitable sources of nitrogen for _A. polymyxa_ (Peoria strain No. 510*). However, without growth factors ammonium sulfate, urea and potassium nitrate were not utilized, nor was a vitamin-free casein hydrolysate a suitable source of nitrogen, when added to a pure salt and glucose medium. In previous work it was found that wheat germ is a suitable source of nitrogen when added to a starch medium (7); that small quantities of urea added to a wheat germ supplemented starch medium increase the glycol yield; and that other protein-containing fractions of the wheat kernel used as nutritional amendments to pure starch, e.g. bran, shorts, feed and white flours, also increase the yield of glycol (13). Bran and shorts used singly (13) were inferior to wheat germ as a supplement to starch medium, whereas white feed flours were somewhat better than bran and shorts but not superior to wheat germ at the concentrations used. Stanier _et al_ (14) report slow fermentation of white flour by _A. polymyxa_ but some increase in rate of fermentation and yield of glycol was obtained when this medium was supplemented with yeast

* N.R.R.L.

extract. It would appear that the A. polymyxa strains so far studied require a proteinaceous nitrogen source to support the fermentation of starch. In order to determine the nitrogen requirements of a number of A. polymyxa strains and other glycol-producing isolates the following tests were carried out.

Nitrogen requirements of A. polymyxa for the fermentation of starch

Introduction

Although wheat germ has been shown (13) to be a suitable nutritional amendment to starch for fermentation by A. polymyxa, at the concentration required to produce maximum yields of glycol this material could not be used economically on a commercial scale.

In order to obtain some general knowledge of the nitrogen requirements of A. polymyxa it was decided to determine the ability of a number of isolates to utilize urea and ammonium sulfate, as well as wheat gluten and starch wash-liquor. The effect of growth factors in combination with the various nitrogen sources was also studied.

Materials and methods

Description of wheat-flour fractions used as supplements to starch

Liquid gluten

It was found in the course of these studies that wheat gluten freed of starch by washing, and stored at low temperatures (10° C) to prevent destructive bacterial growth became quite liquid after a period of two to three weeks. This liquid gluten could be dispensed by the use of a pipette or graduated cylinder, and by shaking in water a suspension of the material was easily prepared. Gluten prepared in this way did not contain a detectable amount of reducing sugar when tested with Benedict's solution. Nitrogen analysis by the Kjeldahl method showed a concentration of 35 to 45 mg. per c.c. for different batches. Gluten prepared from freshly milled flour and from new wheat liquefied more readily than that prepared from old flour. Excessive washing lengthened the time necessary for liquefaction to take place.

Starch wash-liquor

Starch wash-liquor used in these studies was obtained in the process of washing starch from wheat

gluten, as carried out in the laboratory. Flour was mixed with sufficient water to produce a thick dough. After allowing this to stand for one and one-half to two minutes, the starch was washed by hand from the gluten with distilled water. The starch was obtained in a water-suspension from which after a period of a few hours the former separated by sedimentation. The starch-free liquid remaining after sedimentation was removed by decantation and stored at low temperatures for future use. This liquid will be referred to as starch wash-liquor.

Where three litres of water were used in washing the starch from 1000 gm. of flour the nitrogen concentration in the starch wash-liquor was found to be on the average 0.7 mg. per c.c. of liquid (nitrogen determined by the Kjeldahl method).

This liquid has been observed to possess quite marked starch liquefying power when added to gelatinized starch at 60° to 70° C. Furthermore, the liquefied starch appears to be less satisfactory for the Aerobacillus fermentation than the non-liquefied, cooked starch. Hence precautions must be taken to prevent this action of the starch wash-liquor. Liquefaction is prevented by adding the wash-liquor after preliminary cooking of the starch-water suspension at 80° C and by maintaining the medium at this

temperature for 10 to 15 minutes after addition of the wash-liquor.

The studies to be reported were carried out on a 6 percent starch medium. Starch at this concentration produces a medium which, after cooking, can be dispensed into flasks or test-tubes with comparative ease. Higher concentrations were found to be too thick for ease of handling and hence could not have been used without some such treatment as liquefaction by the use of malt or malt extract, which probably would have interfered with the interpretation of the results. Where liquid gluten or starch-wash-liquor was used sufficient distilled water was added to bring the concentration of starch in the medium to 6 percent. All media contained 0.25 percent $CaCO_3$ and 6 percent starch. This will be referred to as "basal" medium.

Media not containing wash-liquor were prepared by cooking the complete medium (basal + nitrogen source) for 20 minutes at 80° C in a steam bath. Where wash-liquor was used, the precautions referred to in the preceding section were observed. The cooked medium was dispensed into test-tubes in 12 to 15 c.c. lots. The tubes were bunged and autoclaved at 15 lb. steam pressure for 20 minutes. Inoculum was prepared by suspending a 5 mm. loopful of bacteria obtained from a 24 to 36 hour old slant in 10 c.c. of sterile 0.9 percent NaCl solution. One similar-sized loopful of the suspension was added to each tube of medium and the inocu-

ed tubes were incubated for the first 24 hours at 35°C, n transferred to a cabinet held at 30° C for the remain- of the fermentation time. At the end of 72 and 96 hours fermentation the extent of starch hydrolysis was determined the use of Lugol's iodine, as an indicator for the presence starch.

It was considered that the qualitative suitability a starch medium could be determined by the complete rolysis of the starch when inoculated with A. polymyxa ains. This procedure was decided upon as suitable for liminary studies with relatively large numbers of anisms, keeping in mind that a medium in which complete rolysis of the starch was obtained would be worthy of ther study, but one in which this was not obtained would unsuitable for the fermentation from an industrial viewpoint.

ults

The reaction of a number of organisms to various rogen-containing compounds is given in Table VII. The ults are averages of three similar experiments.

Complete hydrolysis of the starch was obtained in presence of liquid gluten by all the strains of A. polymyxa ted.

Under the same conditions ammonium sulfate and

lated tubes were incubated for the first 24 hours at 35°C, then transferred to a cabinet held at 30° C for the remainder of the fermentation time. At the end of 72 and 96 hours of fermentation the extent of starch hydrolysis was determined by the use of Lugol's iodine, as an indicator for t e presence of starch.

It was considered that the qualitative suitability of a starch medium could be determined by the complete hydrolysis of the starch when inoculated with A. polymyxa strains. This procedure was decided upon as suitable for preliminary studies with relatively large numbers of organisms, keeping in mind that a medium in which complete hydrolysis of the starch was obtained would be worthy of further study, but one in which this was not obtained would be unsuitable for the fermentation from an industrial viewpoint.

Results

The reaction of a number of organisms to v ious

No growth factors added													
Check (no N. source)		S[+]	S	S	S	S	S	S	S	S	S	S	S
Urea	0.1 gm.	S	S	S	S	S	S	S	S	S	S	S	S
$(NH_4)_2SO_4$	0.2 gm.	S	S	S	S	S	S	S	S	S	S	S	S
Liquid gluten	5 c.c.	S	S	S	S	S	S	0[++]	0	0	0	0	0
Growth factors added*													
Check (no N. source)		S	S	S	S	S	S	S	S	S	S	S	S
Urea	0.1 gm.	S	S	S	S	S	S	S	S	S	S	S	S
$(NH_4)_2SO_4$	0.2 gm.	S	S	S	S	S	S	S	S	S	S	S	S
Liquid gluten	5 c.c.	S	S	S	S	S	S	0	0	0	0	0	0
Yeast extract added**													
Check		S	S	S	S	S	S	S	S	S	S	S	S
Urea	0.1 gm.	S	S	S	S	S	S	S	S	S	S	S	S
$(NH_4)_2SO_4$	0.2 gm.	S	S	S	S	S	S	S	S	S	S	S	S
Liquid gluten	5 c.c.	S	S	S	S	S	S	0	0	0	0	0	0

* Biotin 0.3 micrograms per litre
Thiamin 200 " " "
Nicotinic acid 200 " " "

** 1 gram per litre

\+ S = Starch present as indicated by Lugol's iodine solution

++ 0 =No starch present as indicated by Lugol's iodine solution

urea were not satisfactory nitrogen sources.

The control tubes containing starch wash-liquor were not fermented by A. polymyxa.

Addition of biotin, thiamin and nicotinic acid had no effect on the rate of fermentation of starch in these tests, nor was yeast extract beneficial in this respect.

The Amino Acid Requirements of A. polymyxa

Introduction

The amino acid requirement of many species of bacteria has been determined. Studies of this kind have dealt with a variety of bacteria including soil-borne species as well as several species used in commercial fermentations.

It has been shown by Weizmann (19) that Clostridium acetobutylicum requires asparagine as a nitrogen source and will not grow without biotin in the medium. Andersen and co-workers (1) demonstrated that three species of Lactobacillus, namely L. mannitopeous, L. buckneri and L. lycopersici, all require the amino acids cystine and threonine. In addition tryptophane was found essential for L. buckneri.

West and Lochhead (20) found that 44 percent of the bacteria which they isolated from the rhizosphere of

cultivated plants required some combination of amino acids for their growth. Furthermore biotin and thiamin were the most generally essential growth factors.

Since A. polymyxa is widely distributed in soil (12, 7) and responds generally to biotin and thiamin the results obtained by West and Lochhead (20) appeared to be particularly suggestive. It seemed that this organism might have a specific amino acid requirement. It was decided to investigate this possibility.

Method

The method consisted of adding amino acids singly to the synthetic medium used in the growth factor studies and observing growth after inoculation and incubation, as described in the section dealing with this work.

Glutamic acid was replaced by each of the amino acids referred to in the results. All amino acids were added in the amounts used by West and Lochhead (20), namely 0.05 gm. per litre of medium.

A number of growth factors were added in order that there should exist in the medium a growth-factor supply more similar to that present in wheat than would be the case were biotin and thiamin alone supplied. Wheat has been shown (2) to contain the following growth factors: nicotinic acid, biotin, pantothenic acid, inositol and

thiamin, as well as others. Growth factors were added in the amounts used in the study of the requirement of these substances by A. polymyxa reported in a previous section of this paper. The following factors were added to the "basal" medium, namely biotin, thiamin, nicotinic acid, pantothenic acid and inositol.

The utilization of amino acids by these strains in a reduced growth-factor medium was then studied. Biotin and thiamin were added to the medium. The amino acids added were those which supported the most luxuriant growth in the more complete growth-factor medium.

Results

Amino acid requirements of A. polymyxa in the presence of biotin, thiamin, nicotinic acid, pantothenic acid and inositol

Results obtained in these studies are summarized in Table VIII. Each growth rating is the average of two tests conducted in triplicate.

It is apparent that A. polymyxa exhibits no very specific amino acid requirements, being capable of making growth to a greater or lesser extent on all but five of eighteen amino acids included in the study.

Marked and abundant growth was obtained where the following amino acids forming the sole source of nitrogen

TABLE VIII

The amino acid requirement of different strains of <u>A. polymyxa</u> in the presence of biotin, thiamin, pantothenic acid, nicotinic acid and inositol

	Growth rating of strains used				
Amino acid	C42(2)	C47(2)	C56(2)	643	641
dl-alpha-alanine	5	5	5	4	4
d-glutamic acid	4	4	5	5	4
l-aspartic acid	5	4	5	4	5
dl-serine	4	4	4	4	4
l-leucine	3	4	4	4	3
l-cystine	3	2	4	2	2
l-cysteine	3	2	3	3	3
d-arginine	1	2	3	1	3
dl-phenylalanine	2	1	1	2	2
dl-methionine	1	1	2	2	2
dl-valine	2	3	3	2	2
dl-isoleucine	2	2	2	1	3
d-lysine	0	0	0	0	0
l-histidine	0	0	0	0	0
l-tryptophane	0	0	0	0	0
l-proline	0	0	0	0	0
l-tyrosine	0	0	0	0	0
Check (no amino acid)	0	0	0	0	0

were added singly to the medium: dl-alpha-alanine, d-glutamic acid, l-aspartic acid, dl-serine and l-leucine. Single acids of another group supported meagre or inconsistent growth. Included in this group were l-cystine, l-cysteine, d-arginine, dl-phenylalanine, dl-methionine, dl-valine and dl-isoleucine. The amino acids d-lysine, l-histidine, l-tryptophane, l-proline and l-tyrosine were generally

unsuitable for any of the <u>A. polymyxa</u> strains studied.

No marked difference in amino acid requirements between the various strains of <u>A. polymyxa</u> used were noted in these experiments.

<u>Amino acid requirements in the presence of biotin and thiamin</u>

A summary of the studies included under this heading is given in Table IX.

TABLE IX

The amino acid requirement of <u>A. polymyxa</u> in the presence of biotin and thiamin

Amino acid	Growth rating of strains used				
	C42(2)	C47(2)	C56(2)	643	641
dl-alpha-alanine	3	0	4	4	3
d-glutamic acid	5	4	4	5	4
dl-serine	4	tr*	3	4	3
l-aspartic acid	5	3	4	4	4
l-cystine	3	0	3	3	2
All amino acids	5	5	5	5	5
Check (no amino acids)	0	0	0	0	0
All amino acids (no growth factors)	0	0	0	0	0

* Trace

From Table IX it may be noted that generally the restricted growth-factor supply somewhat reduced the

growth of A. polymyxa compared with that obtained in the presence of the more complete factor environment.

Strain C47(2) showed most reduction in growth as a result of removal of nicotinic acid, pantothenic acid and inositol. No growth of this organism was obtained in medium containing alanine or cystine and only a trace in the presence of serine.

Growth of all strains in the presence of thiamin, biotin and a mixture of the amino acids included in the study, was uniformly abundant. However no growth was observed when the growth factors were withdrawn.

Inorganic Nitrogen and Urea as Nitrogen Sources for A. polymyxa

Introduction

The non-specific nature of the amino acid requirement of A. polymyxa suggested the possibility that some nitrogen-containing substances other than protein or protein derivatives might be utilized by this organism. A test was therefore conducted to determine the suitability of ammonium sulfate, potassium nitrate and urea in this regard.

Method

The above named salts were used singly to replace the amino acids in the basal synthetic medium used in the previous studies on nitrogen requirements of A. polymyxa. The usual amounts of biotin, thiamin, nicotinic acid, pantothenic acid and inositol were used. The procedure followed was that described in the section dealing with growth-factor requirements.

All growth ratings were made following two successive transfers in the medium.

Results

A summary of the results obtained in two consecutive trials conducted in duplicate is presented in Table X.

TABLE X

The utilization of inorganic nitrogen and urea by A. polymyxa

Nitrogen source	Conc. in gm. per l.	Growth rating of strain used					
		C42(2)	C47(2)	C56(2)	UA634	UA641	UA652
$(NH_4)_2SO_4$	0.1	4	0	0	0	0	4
KNO_3	0.1	0	0	0	0	0	0
Urea	0.1	0	0	0	0	0	0
Check		0	0	0	0	0	0

Two strains of A. polymyxa, namely C42(2) and U.A.652 showed consistent ability to utilize ammonium sulfate as a source of nitrogen. Although some growth was occasionally made at first by other strains, no growth was recorded after the first transfer in this medium.

No strains produced growth with urea or potassium nitrate as sources of nitrogen.

Training of A. polymyxa to Utilize Ammonium Sulfate

Introduction

It was noticed in the course of previous work that preculturing A. polymyxa on a glucose-salts-growth-factor medium with glutamic acid as a source of nitrogen had some effect upon the ability of these bacteria to utilize ammonium sulfate. Although most of the literature on the subject of training deals with organisms exhibiting some specific amino acid requirement, reference to some of this work may be of value.

According to Gladstone (6) Knight defines training as the "derivation of cultures having simpler nutrient requirements from cultures with a more complex nutrition."

Fildes et al (3) states that strains of Bact. typhosum normally requiring specific amino acids,

especially tryptophane, could be trained to dispense with this substance and utilize ammonia as a source of nitrogen. These workers also demonstrated that at least part of the training to dispense with a supply of tryptophane was due to acquired ability to synthesize this essential amino acid. A recent review of this subject is given by Gale (5).

Since, as has been shown in this paper, _A. polymyxa_ does not appear to have specific amino acid requirements, it must be capable of synthesizing all the necessary amino acids which go into the make-up of its protein and enzymes. This has been shown in the case of strains C42(2) and U.A.652 which utilize ammonium sulfate as a sole source of nitrogen.

Training of a number of strains to utilize ammonium sulfate should be a contribution to the development of a process for the conversion of pure starch to glycol.

Method

Generally the method consisted of preculturing a number of strains in glutamic acid medium and comparing the ability of normal and "trained" strains to utilize ammonium sulfate as a source of nitrogen.

Basal medium containing biotin and thiamin in the usual amounts, as described in the section dealing with growth-factor requirements, was used throughout these

tests with the exception of controls in which growth factors were omitted.

The training medium contained 0.05 gm. glutamic acid per litre as a source of nitrogen. The medium used for comparison of "trained" and normal strains contained 0.1 gm. $(NH_4)_2SO_4$ per litre.

The procedure followed for the inoculation of the training medium and the ammonium sulfate medium with the normal trains was that described for the experiments referred to above. Inoculum containing the "trained" strains was prepared by adding two loopsful of the training medium in which the bacteria had grown for 72 hours to 10 c.c. of sterile 0.9 percent NaCl solution.

Growth in the ammonium sulfate medium was recorded after incubation for 48 hours at 35° C. The results given are those observed in the third generation in this medium.

Results

Table XI contains a summary of the average results obtained in two tests.

All "trained" isolates grew well in the medium containing ammonium sulfate as a source of nitrogen and in the presence of biotin and thiamin. No growth was obtained in the absence of the growth factors.

Five of the nine strains tested failed to make

TABLE XI

Training effect of glutamic acid on A. polymyxa

Treatment of Strain	Medium used	Growth rating of strains used								
		C42(2)	C47(2)	C56(2)	U.A.634	U.A.641	U.A.645	U.A.650	U.A.651	U.A.652
Normal										
	$(NH_4)_2SO_4$ + G.F.*	4	0	tr+	0	0	0	tr	0	5
	$(NH_4)_2SO_4$	0	0	0	0	0	0	0	0	0
Trained										
	$(NH_4)_2SO_4$ + G.F.*	5	5	5	5	5	5	5	5	3
	$(NH_4)_2SO_4$	0	0	0	0	0	0	0	0	0

* Growth factor
\+ Trace

growth in the ammonium sulfate medium without training. Two of the remaining four made slight and irregular growth under these conditions.

As shown in previous tests C42(2) and U.A.652 utilized ammonium sulfate without training.

The Effect of Liquid Gluten on the Glycol and Ethanol Yields of *A. polymyxa* Starch Fermentations

Introduction

The effect of the concentration of nitrogen upon the yield of products obtained has been studied for a number of fermentation processes. Studying the effect of carbohydrate-protein ratio upon the yields of products in the acetone-butanol fermentation, Fulton *et al* (4) showed that a low C:N ratio favors increased yields of acetone and decreased yields of ethanol. A high C:N ratio has the opposite effect upon the yield of solvents. Van Niel (18) found that the propionic acid - acetic acid ratio in the propionic acid fermentation was influenced by the concentration of nitrogenous material in the medium. A high concentration of peptone was shown to be associated with a high propionic acid - acetic acid ratio, whereas a low concentration of peptone was associated with a low ratio.

Having found gluten to be a suitable source of nitrogen for the *Aerobacillus* fermentation of starch it was decided to investigate the effect of gluten concentration on the yield of products formed.

Methods

The medium used contained the following substances

at the concentrations given per 100 c.c. of liquids and water: starch, 6 gm.; $CaCO_3$, 0.25 gm.; starch wash-liquor, 10 c.c.; liquid gluten in different concentrations. The latter, when added, was estimated to contain approximately 50 percent water. Gluten concentration varied as indicated on the graph in the following section. The starch, liquid gluten, and water mixture was cooked for 10 minutes in a steam bath at the temperature of boiling water. At the end of this period starch wash-liquor was added and cooking was continued for a further period of 15 minutes. $CaCO_3$ was then added and dispensed throughout the medium by thorough stirring.

This medium was dispensed into 250 c.c. flasks in 150 c.c. lots. The flasks were then bunged and autoclaved at 15 lb. pressure for 30 minutes.

Inoculum was prepared by transferring the bacterial growth from a 24- to 36-hour-old agar slant to 100 c.c. of sterile water. The contents of the flask were then shaken cautiously to bring about uniform dispersion of the cells in the liquid. This suspension was immediately added at the rate of 5 c.c. per 150 c.c. of medium, and distribution of the bacterial cells was obtained by shaking the inoculated flasks. Fermentation was carried out at 35° C for the first 8 to 12 hours, and at 30° C for the remainder of the fermentation period. This fermentation was found to be somewhat slower than that of whole wheat,

and consequently 96-hour periods were used instead of the usual 72 hours used for wheat.

Results

The general effect of gluten on glycol and alcohol yield is shown by the accompanying graph (Figure 4). Each point represents an average of two tests conducted in duplicate.

Concentration of gluten in the medium had an effect on the glycol - alcohol ratio as shown in the accompanying figure.

Increasing gluten concentration up to 600 mg. of nitrogen per 100 c.c. of medium caused a continued increase in glycol yield.

The addition of 0.5 to 1 c.c. of liquid gluten per 100 c.c. medium caused alcohol to reach a concentration of just under 1 percent. This did not increase with further additions of gluten up to 12 c.c. per 100 c.c. of medium.

The Effect of Ammonium Sulfate at Various Gluten Concentrations on Glycol Yields From Starch Fermentations

Introduction

In many of the tests carried out ammonium sulfate

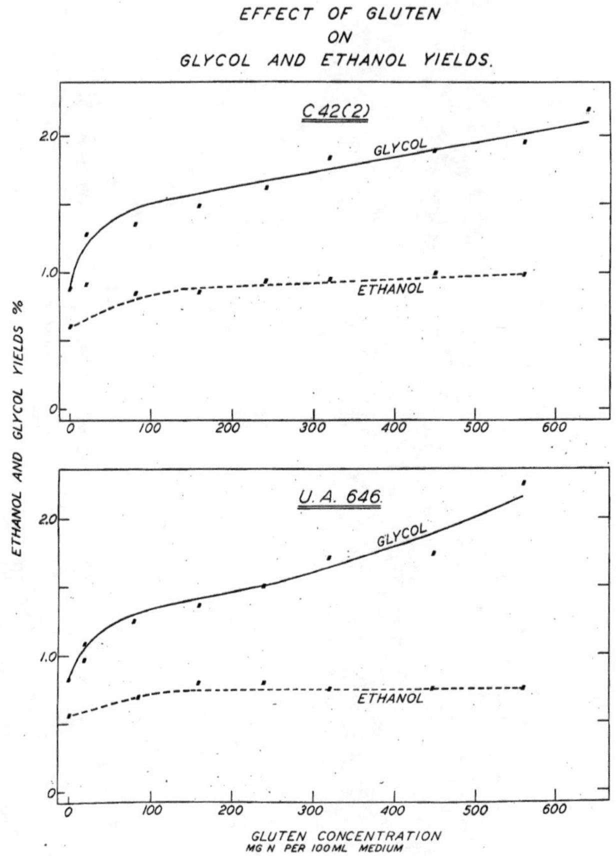
EFFECT OF GLUTEN
ON
GLYCOL AND ETHANOL YIELDS.
C 42(2)
GLYCOL
ETHANOL
U. A. 646.
GLYCOL
ETHANOL
ETHANOL AND GLYCOL YIELDS %
2.0
1.0
0
0
100
200
300
400
500
600
GLUTEN CONCENTRATION
MG N PER 100 ML MEDIUM

Figure 4

appeared to have some effect on starch fermentations by A. polymyxa. To determine the effect upon yield of glycol the following test was conducted.

Method

The media used were similar to those in the previous experiment, and were prepared in the same manner. Liquid gluten and ammonium sulfate were added in the amounts indicated in the tabulation of results. Inoculum was prepared by suspending the bacteria in sterile water and the conditions of temperature for incubation were followed as in the previous experiment. Analysis for glycol and alcohol were made after 96 hours' fermentation.

Results

Table XII is a summary of the results obtained, with the phage-resistant strain C42(2) and the high-glycol-yielding strain U.A.634.

Glycol yields obtained from these strains indicate that ammonium sulfate is utilized by both.

The optimum ammonium sulfate concentration shows a tendency to decrease as the gluten concentration increases.

Generally higher yields are obtained where gluten plus ammonium sulfate at 0.01 percent is used, rather than gluten alone.

TABLE XII

Effect of ammonium sulfate on the glycol yield of A. polymyxa strains C42(2) and U.A.634

% $(NH_4)_2SO_4$	% Glycol yield from strains used							
	0.0*		0.5		1.0		2.0	
	C42(2)	634	C42(2)	634	C42(2)	634	C42(2)	634
0.0	0.28	0.73	1.10	1.26	1.21	1.52	1.31	1.71
0.01	0.32	0.88	1.24	1.33	1.44	1.76	1.58	1.67
0.05	0.60	0.96	1.09	1.20	1.21	1.48	1.39	1.59
0.1	0.86	1.13	1.05	1.26	1.11	1.44	1.28	1.58
0.15	0.72	1.09	0.99	1.35	1.11	1.35	1.22	1.54

* c.c. of liquid gluten added per 100 ml. of medium

Isolate U.A.634 which has been shown to out-yield C42(2) on wheat mashes (Table IV) does so consistently in the starch medium containing gluten and ammonium sulfate.

Discussion

Although ammonium sulfate alone does not appear to be a satisfactory source of nitrogen for A. polymyxa in wheat starch fermentations, yet glycol analyses of the fermenting media show that this nitrogen source is utilized to some extent by this organism. Furthermore, in the presence of gluten increased yields of glycol are obtained in starch fermentations from the addition of small amounts of ammonium sulfate. However, the optimum ammonium sulfate

is lower where gluten is added to the medium than where it is absent. This indicates that under the conditions of these experiments gluten is the preferred source of nitrogen for <u>A. polymyxa</u>.

Since both C42(2) and U.A.634 respond to ammonium sulfate in the starch fermentations and only the former has been shown to consistently utilize this source in studies where it formed the only source of nitrogen, some modifications of the latter strain must take place when grown in starch medium containing at least traces of amino nitrogen, which enables it to utilize ammonia nitrogen. That a change of this kind may take place has been shown in the study on the training of <u>A. polymyxa</u> to utilize ammonium sulfate. Such behavior might explain the increases in glycol yield obtained where ammonium sulfate is added to the starch medium.

SUMMARY

1. Pasteurization of strains of *A. polymyxa* can be used to increase their glycol-yielding ability in the fermentation of 8 percent wheat mashes.

2. Pasteurization of source material was shown to be an aid in the isolation of high-glycol-yielding strains of *A. polymyxa*.

3. Soil and rotting wood were found to be good sources of *A. polymyxa* isolates.

4. Strains of *A. polymyxa* from Alberta sources compared favorably in glycol production with those isolated elsewhere.

5. A large proportion of the high-glycol-yielding isolates of *A. polymyxa* obtained in the pasteurization studies fell into a single group based on dye absorption and colony characters.

6. The strains studied do not form spores on glucose-agar medium.

7. All strains grown on non-cooked potato slices rotted this material in less than 72 hours at 30° C.

8. Biotin was the only growth factor found essential

for all strains tested.

9. Improved growth over that obtained with biotin alone was obtained where biotin plus any one of the following growth factors was present in the medium: thiamin, pantothenic acid, nicotinic acid or inositol.

10. Optimum growth was obtained when a mixture of biotin and thiamin plus one of the following was present in the medium: pantothenic acid, nicotinic acid or inositol.

11. A form of liquid gluten apparently formed by enzymatic action at low temperature proved to be one of the most suitable sources of nitrogen for A. polymyxa in the fermentation of wheat starch.

12. Under the same conditions ammonium sulfate and urea did not support complete hydrolysis of the starch by A. polymyxa.

13. The addition of the growth factors biotin, nicotinic acid and thiamin to the above media did not improve growth in the presence of either gluten, ammonium sulfate or urea.

14. A. polymyxa does not exhibit any very specific amino acid requirements.

15. In a medium containing glucose, salts, and the growth factors biotin, thiamin, nicotinic acid, pantothenic acid and inositol, good growth was made by <u>A. polymyxa</u> where the following amino acids, used singly, were the source of nitrogen: glutamic acid, aspartic acid, alanine, serine and leucine.

16. Under these conditions meagre and inconsistent growth was obtained with cystine, cysteine, phenylalanine, methionine and isoleucine.

17. The following amino acids were not utilized by <u>A. polymyxa</u>: lysine, tryptophane, histidine, proline and tyrosine.

18. Amino acids supporting good growth in the above medium did so in the presence of a reduced growth-factor supply including biotin and thiamin only.

19. In the complete growth-factor medium in which the amino acids were replaced by ammonium sulfate, potassium nitrate and urea, none of the strains tested utilized the latter two sources of nitrogen, and only two strains, namely C42(2) and U.A.652, utilized ammonium sulfate.

20. A number of strains of <u>A. polymyxa</u> were trained to utilize ammonium sulfate by preculturing in a medium containing glutamic acid as the sole source of nitrogen.

21. Within the range studied, increasing protein nitrogen in the form of liquid gluten increased the yields of glycol from wheat starch.

22. The C:N ratio of the medium had an effect upon the ratio of ethanol to glycol. Increasing nitrogen concentrations caused an increased glycol - ethanol ratio.

23. In starch media containing liquid gluten and ammonium sulfate the former source of nitrogen was preferred by A. polymyxa.

ACKNOWLEDGEMENTS

The writer wishes to express his thanks to Dr. A. W. Henry, Associate Professor of Plant Pathology, under whose direction this work has been carried out, for his helpful criticism during the preparation of this manuscript.

Acknowledgement is expressed to W. G. Corns and A. L. Shewfelt who carried out the glycol determinations, and to W. E. Brown and J. Kastelic who conducted the sugar and alcohol determinations. Thanks are also due to G. M. Tosh, Technician, for construction of laboratory equipment.

Financial assistance was supplied by the National Research Council.

LITERATURE CITED

1. ANDERSEN, A.A., WOOD, H.G. and WERKMAN, C.H. Amino acid requirements of the lactic acid bacteria. Jour. Bact. 36:655. 1938.

2. BURKHOLDER, PAUL R. The vitamins in dehydrated seeds and sprouts. Sci. 97:562-564. 1943.

3. FILDES, P., GLADSTONE, G.P. and KNIGHT, B.C.J.G. The nitrogen and vitamin requirements of Bact. typhosus. Br. Jour. Exp. Path. 14:189-196. 1933.

4. FULTON, H.L., PETERSON, W.H. and FRED, E.B. The hydrolysis of native proteins by B. granulobacter pectinovorum and the influence of the carbohydrate-protein ratio on the products of fermentation. Cent. Bakt. Parasitenk. Abt. II. 27:1-11. 1926.

5. GALE, E.F. Factors influencing the enzymic activities of bacteria. Bact. Rev. 7:139-173. 1943.

6. GLADSTONE, G.P. The nutrition of Staphylococcus aureus; nitrogen requirements. Br. Jour. Exp. Path. 18:322-333. 1937.

7. HENRY, A.W. and JACKSON, A.W. Report B. Studies on the production of 2,3-butylene glycol from wheat starch. Canadian Investigations on the production and esterification of 2,3-butylene glycol. Can. Nat. Res. Counc. Rept. Dec., 1942.

8. KATZNELSON, H. Report C. Studies on the nutritional requirements of A. polymyxa. Canadian investigations on the production and esterification of 2,3-butylene glycol. Can. Nat. Res. Counc. Rept. Dec., 1942.

9. LIEBMAN, A.J. Alcohol in War. Wallerstein Lab. Communic. 15:131-140. 1942.

10. National Research Council Rept. D. Studies on the freezing point curves of 2,3-butylene glycol-water mixtures. A.E. Chadderton and W.H. Cook. Canadian investigations on the production and esterification of 2,3-butylene glycol. Dec., 1942.

11. O'MEARA, R.A.Q. A simple, delicate and rapid method of detecting the formation of acetylmethylcarbinol by bacteria fermenting carbohydrates. J. Path. and Bact. 4:401-406. 1931.

12. PORTER, R., McCLESKEY, C.S. and LEVINE, M. The facultative sporulating bacteria producing gas from lactose. Jour. Bact. 33:163-183. 1937.

13. SHEWFELT, A.L. A report on the production of 2,3-butylene glycol from wheat and wheat starch. Unpub. M.Sc. Thesis Univ. of Alta. 1943.

14. STANIER, R.Y., ADAMS, G.A. and LEDINGHAM, G.A. Report A. Studies on factors affecting the Aerobacillus fermentation. Canadian investigations on the production and esterification of 2,3-butylene glycol. Can. Nat. Res. Counc. Rept. Dec., 1942.

15. ____________ CLEMENT, MARY T., DWARKIN, S. and LEDINGHAM, G.A. Report A. Isolation and characteristics of Aerobacillus. Canadian investigations on the production and esterification of 2,3-butylene glycol. Can. Nat. Res. Counc. Rept. Dec., 1942.

16. TILDEN, E.B. and HUDSON, C.S. Preparation and properties of the amylases produced by Bac. macerans and B. polymyxa. Jour. Bact. 43:527-544. 1942.

17. University of Alberta Report No.3. Studies on production of 2,3-butylene glycol from grain. Sept., 1943.

18. VAN NIEL, C.B. The propionic acid bacteria. Thesis Technische Hoogeschool Delft. Sept., 1928.

19. WEIZMANN, C. and ROSENFELD, B. The specific nutrient requirements of Cl. acetobutylicum (Weizmann). Bioch. Jour. 33:1376-1389. 1939.

20. WEST, P.M., LOCHHEAD, A.G. Qualitative studies of soil micro-organisms. IV. The rhizosphere in relation to the nutrient requirements of soil bacteria. Can. J. Res. C18:129-135. 1940.

21. WEYER, E.R. and RETTGER, L.F. A comparative study of six different strains of the organism commonly concerned in large scale production of butyl alcohol and acetone by the biological process. Jour. Bact. 14:399-424. 1927.